NEPTUNE

by Emma Bassier

Cody Koala

An Imprint of Pop!
popbooksonline.com

abdobooks.com
Published by Pop!, a division of ABDO, PO Box 398166, Minneapolis, Minnesota 55439.

Printed in the United States of America, North Mankato, Minnesota.

102020
012021

Cover Photos: NASA, foreground; iStockphoto, background
Interior Photos: NASA, 1 (foreground), 19 (top), 19 (bottom left), 19 (bottom right); iStockphoto, 1 (background), 5, 6, 9 (bottom left), 15; Shutterstock Images, 9 (top), 11, 12, 16 (Sun), 16 (Neptune); Detlev van Ravesnwaay/Science Source, 9 (bottom right); QA International/Science Source, 20

Editor: Alyssa Krekelberg
Series Designer: Colleen McLaren

Library of Congress Control Number: 2020940264
Publisher's Cataloging-in-Publication Data
Names: Bassier, Emma, author.
Title: Neptune / by Emma Bassier
Description: Minneapolis, Minnesota : POP!, 2021 | Series: Planets | Includes online resources and index
Identifiers: ISBN 9781532169113 (lib. bdg.) | ISBN 9781532169472 (ebook)
Subjects: LCSH: Neptune (Planet)--Juvenile literature. | Planets--Juvenile literature. | Solar system—Juvenile literature. | Milky Way--Juvenile literature. | Space--Juvenile literature.
Classification: DDC 523.48--dc23

Hello! My name is

Cody Koala

Pop open this book and you'll find QR codes like this one, loaded with information, so you can learn even more!

Scan this code* and others like it while you read, or visit the website below to make this book pop.

popbooksonline.com/neptune

*Scanning QR codes requires a web-enabled smart device with a QR code reader app and a camera.

Table of Contents

Chapter 1

The Last Planet

There are eight planets
in our **solar system**.
Neptune is the last one.
All planets **orbit** the Sun.
The Sun is in the center
of the solar system.

Neptune
Watch a video here!

Since Neptune is so far away from the Sun, it gets little heat and light. The planet is approximately –328 degrees Fahrenheit (–200°C).

It takes four hours for light from the Sun to reach Neptune.

Chapter 2

Cold and Windy

Neptune is approximately four times as wide as Earth. The huge planet does not have a solid surface. It is made of gas and ice.

Learn more here!

Neptune has several layers. The top layer is the **atmosphere**. It is made of gases and cold clouds. Next comes the mantle. This is a layer of icy materials. At the center of the planet is a heavy, rocky core.

Inside Neptune

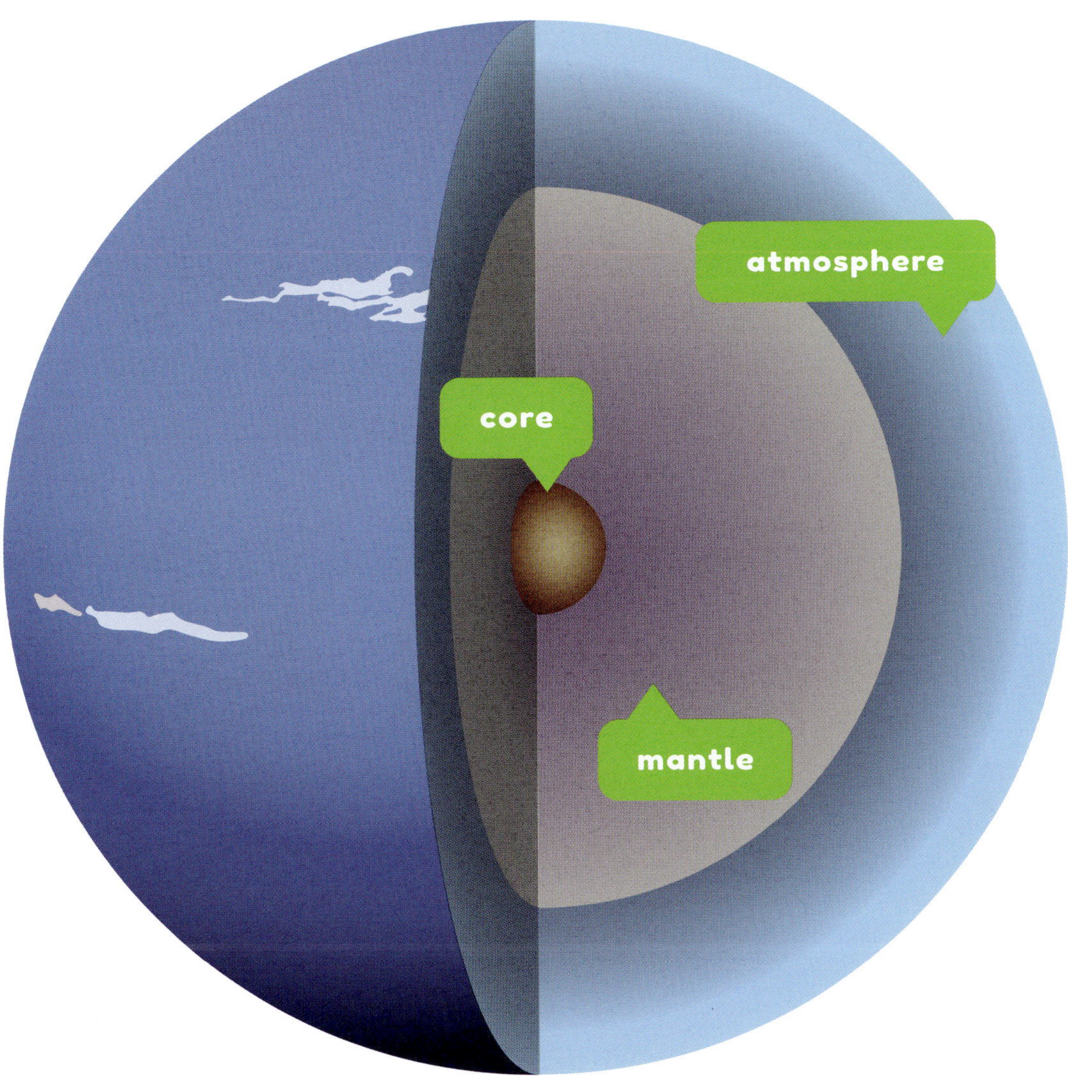

Neptune's winds are much faster than winds on Earth.

Neptune is the windiest planet. Winds move at 1,243 miles per hour (2,000 kmh). That is faster than most airplanes can fly.

Some storms on Neptune are so large they can be seen from space.

Chapter 3

A Very Long Year

Neptune has the longest **orbit** of any planet. One orbit around the Sun is one year. A year on Neptune lasts 165 Earth years.

Complete an activity here!

Neptune

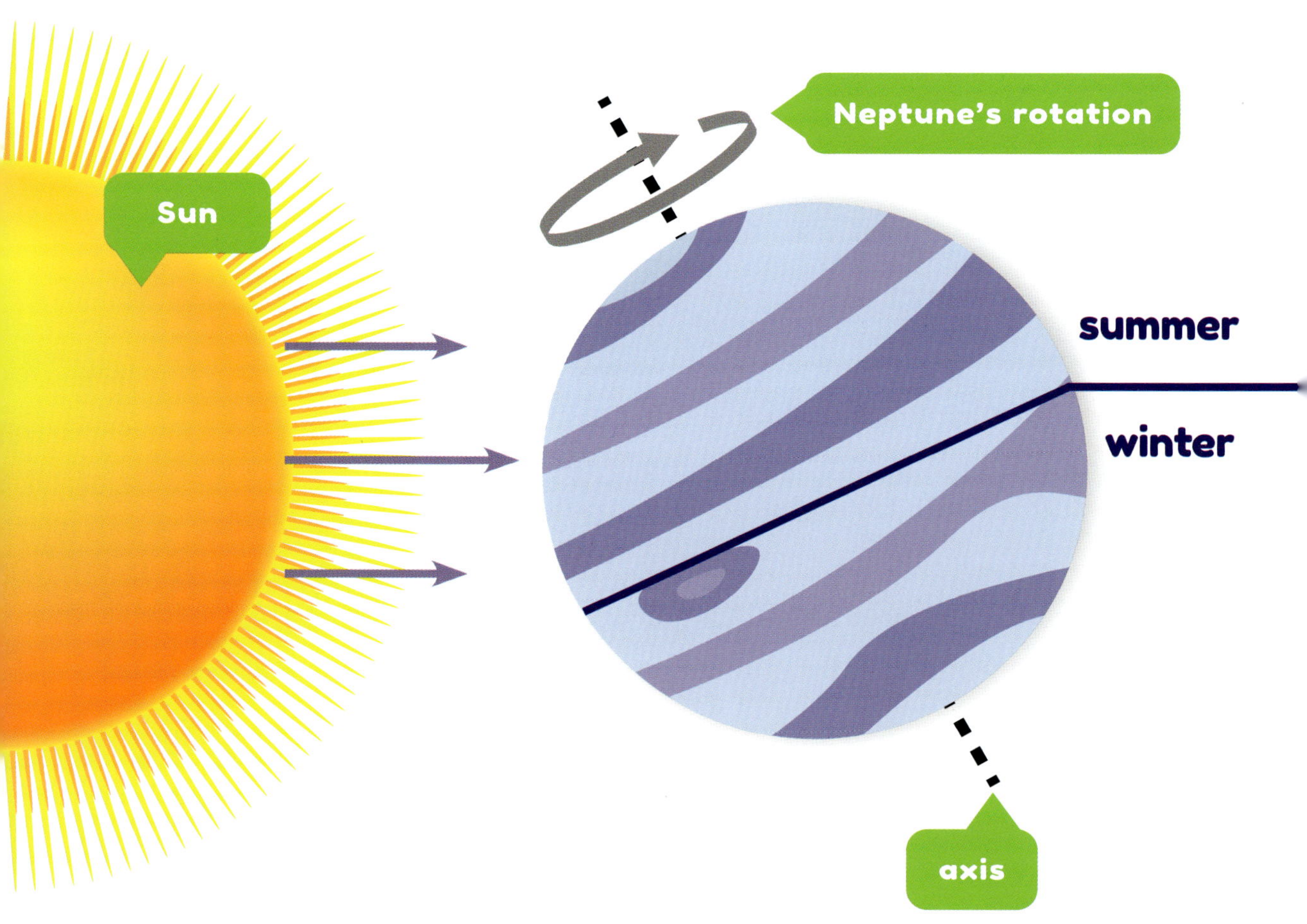

Neptune spins fast. One spin on its **axis** is the length of one day. Its day is 16 hours long. Neptune's axis is tilted. The tilt causes seasons. Each season lasts more than 40 years.

Chapter 4

Moons and Rings

Neptune has 14 moons. The largest moon is Triton. It **orbits** in the opposite direction its planet spins. This is unusual to see in our **solar system**.

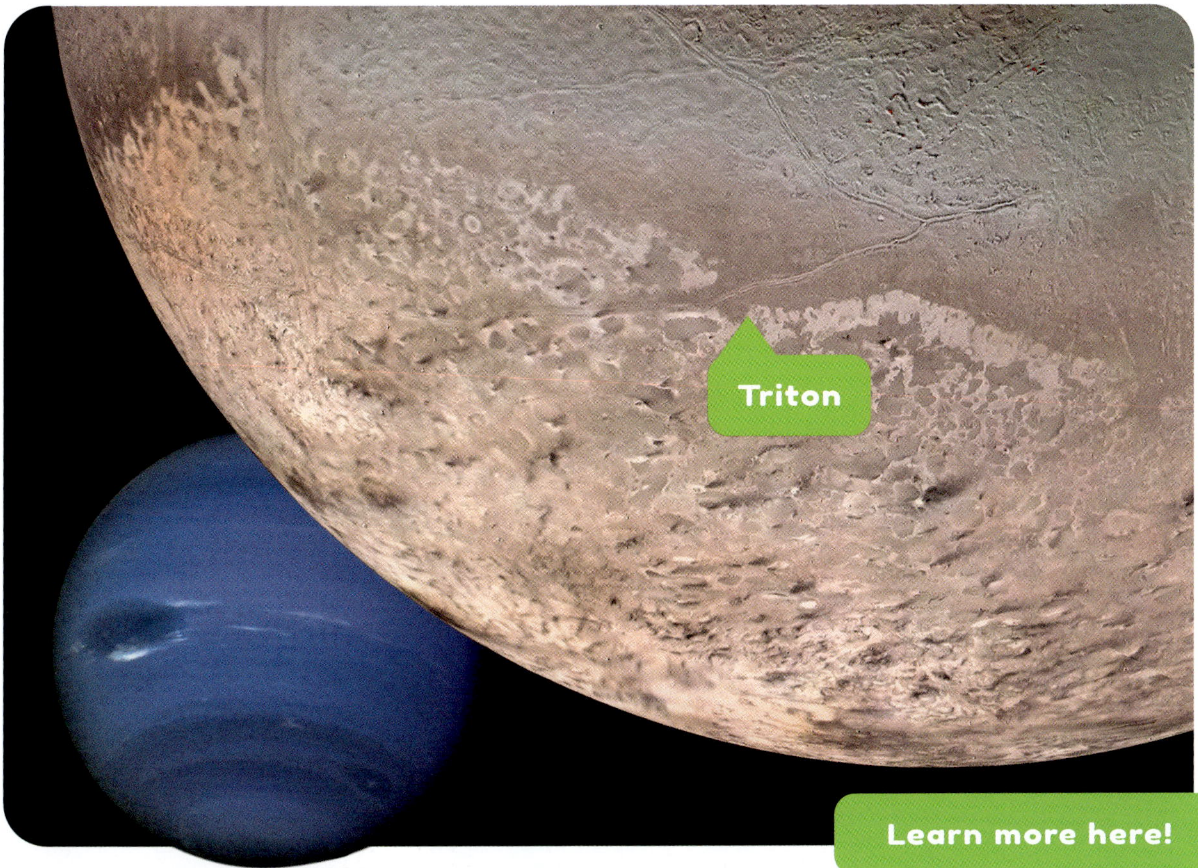

Triton

Learn more here!

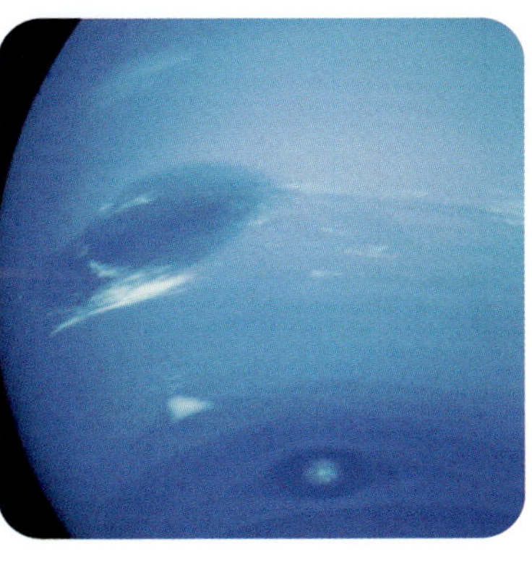

One spacecraft has visited Neptune. It was called *Voyager 2*. It showed that Neptune has at least five **faint** rings. The rings are made of dust and small rock chunks.

Making Connections

Text-to-Self

Each season on Neptune lasts more than 40 years. Would you be happy if Earth's seasons lasted that long? Why or why not?

Text-to-Text

Have you read other books about planets? How are those planets similar to or different from Neptune?

Text-to-World

Living things can survive on Earth but not on Neptune. What are some things Earth has that living things need to survive?

Glossary

atmosphere – the layers of gases that surround a planet.

axis – an imaginary line that runs through the middle of a planet, from top to bottom.

faint – being light or hard to see.

orbit – to follow a rounded path around another object.

solar system – a collection of planets and other space material orbiting a star.

Index

Online Resources

popbooksonline.com

Thanks for reading this Cody Koala book!

Scan this code* and others like it in this book, or visit the website below to make this book pop!

popbooksonline.com/neptune

*Scanning QR codes requires a web-enabled smart device with a QR code reader app and a camera.